Nature's Mysteries

YELLOWSTONE'S BOILING RIVER

by Patricia Hutchison

abdobooks.com

Published by Pop!, a division of ABDO, PO Box 398166, Minneapolis, Minnesota 55439.

Printed in the United States of America, North Mankato, Minnesota.

102020
012021

THIS BOOK CONTAINS RECYCLED MATERIALS

Cover Photo: Shutterstock Images
Interior Photos: Shutterstock Images, 1, 7, 13, 15, 18, 19, 22–23, 25, 29; Dave Walsh/VW Pics/Contributor/Universal Images Group/Getty Images, 5; Erik Petersen/The Livingston Enterprise/AP Images, 6; TMI/Alamy, 8; Norbert Rosing/National Geographic Image Collection/Getty Images, 11; iStockphoto, 17, 20, 21, 26, 27, 28

Editor: Alyssa Krekelberg
Series Designers: Candice Keimig, Victoria Bates, and Laura Graphenteen

Library of Congress Control Number: 2020940273

Publisher's Cataloging-in-Publication Data

Names: Hutchison, Patricia, author.

Title: Yellowstone's boiling river / by Patricia Hutchison

Description: Minneapolis, Minnesota : POP!, 2021 | Series: Nature's mysteries | Includes online resources and index

Identifiers: ISBN 9781532169236 (lib. bdg.) | ISBN 9781532169595 (ebook)

Subjects: LCSH: Yellowstone National Park--Juvenile literature. | Hot springs--Juvenile literature. | Curiosities and wonders--Juvenile literature. | Mystery--Juvenile literature. | Geography--Juvenile literature.

Classification: DDC 910.02--dc23

WELCOME TO DiscoverRoo!

Pop open this book and you'll find QR codes loaded with information, so you can learn even more!

Scan this code* and others like it while you read, or visit the website below to make this book pop!

popbooksonline.com/boiling-river

*Scanning QR codes requires a web-enabled smart device with a QR code reader app and a camera.

TABLE OF CONTENTS

CHAPTER 1

NATURE'S HOT TUB

Steam rises from an area where two rivers meet. Hikers walk carefully into the water. One person thinks the water is too warm. A little farther away, another person thinks the water is too cold.

WATCH A VIDEO HERE!

They meet in the middle. The water temperature there is perfect.

People of all ages visit the natural hot tub in Yellowstone National Park.

Animals such as elk also visit the natural hot tub.

This natural hot tub is in Yellowstone National Park. The water is heated by the Boiling River. This river is so hot that it can burn people's skin. But at the hot tub, the hot water mixes with cool water from the Gardner River. This spot is perfect for soaking or swimming.

WHERE IS YELLOWSTONE NATIONAL PARK?

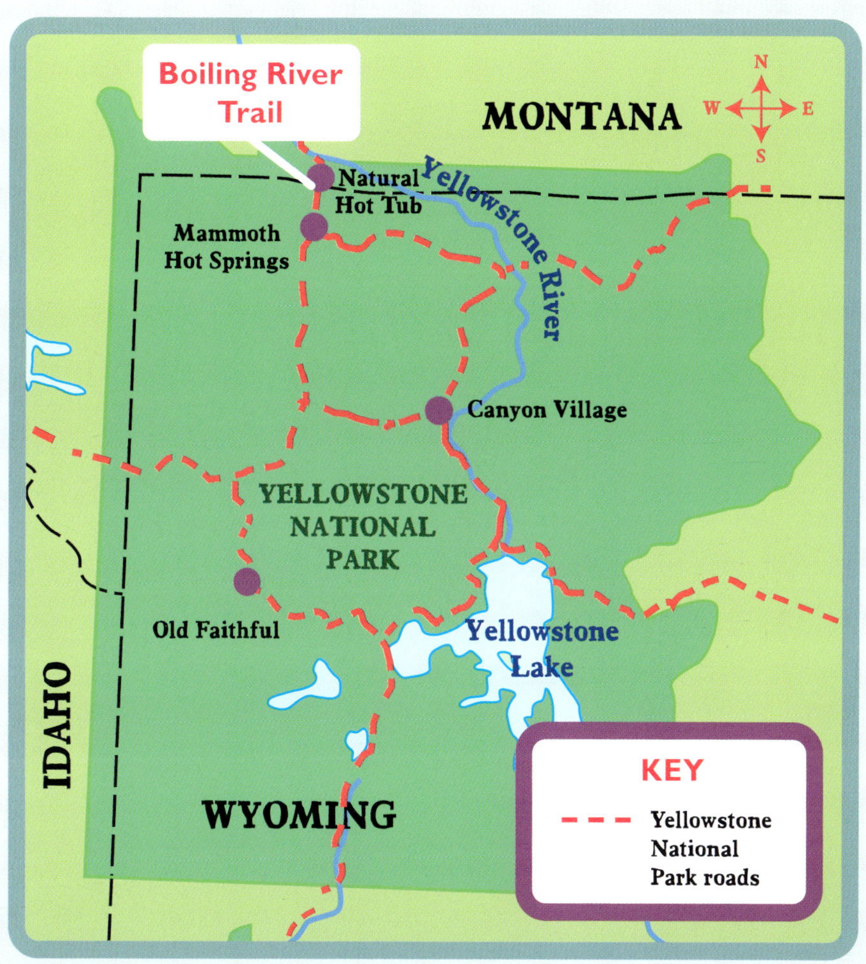

Yellowstone National Park is mostly in Wyoming. Parts of it cross into Montana and Idaho.

The natural hot tub near the Boiling River is open in the summer, fall, and winter.

People can visit Yellowstone to soak in the hot tub. They can relax in the warm water, even when it starts to get cold outside. Scientists come to Yellowstone too. They study its unique features, such as the Boiling River.

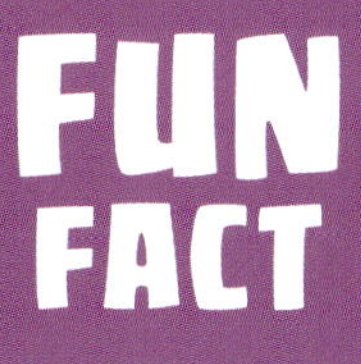

More than four million people visit Yellowstone National Park each year.

CHAPTER 2

SMALL STREAM, BIG EFFECT

The Boiling River is really just a stream. It is only about 6.5 feet (2 m) wide. It is 100 yards (91 m) long. The water is nearly 140 degrees Fahrenheit (60°C) as it streams out of the ground. It flows into

LEARN MORE HERE!

The Boiling River's water isn't actually boiling. But it is incredibly hot and dangerous to touch.

the much colder water of the Gardner River. This creates clouds of steam.

The hot and cold waters swirl together. A person could have one foot in very hot water. The other foot could be in ice-cold water. And sometimes water currents change quickly. A person's swimming spot might suddenly get too cold or hot.

CLOSED IN THE SPRING

The swimming area near the Boiling River is closed in the spring. Water from melting snow rushes into the Gardner River. The river becomes fast and dangerous. Lots of cold water pours into the natural hot tub. The water becomes too high and chilly for people to swim safely.

Hikers go along a trail and over a bridge to get to the natural hot tub.

FUN FACT

Some visitors do a polar plunge. First, they sit in the hot water. Then, they jump into the cold river to cool off.

CHAPTER 3

HOT SPRINGS AND A VOLCANO

Scientists think Boiling River's hot water comes from Mammoth Hot Springs. This area is approximately 2 miles (3.2 km) south of the Boiling River. Yellowstone has many hot springs. That's because the park is on a **caldera**.

LEARN MORE HERE!

Mammoth Hot Springs has approximately 50 hot springs.

The caldera was created by a **supervolcano** that erupted long ago. Scientists studied the area to learn about this volcano and what it left behind. They think it formed millions of years ago.

Yellowstone's supervolcano is still active. But scientists don't think it will erupt anytime soon.

The supervolcano's eruption caused the ground beneath it to collapse. This is what made the caldera.

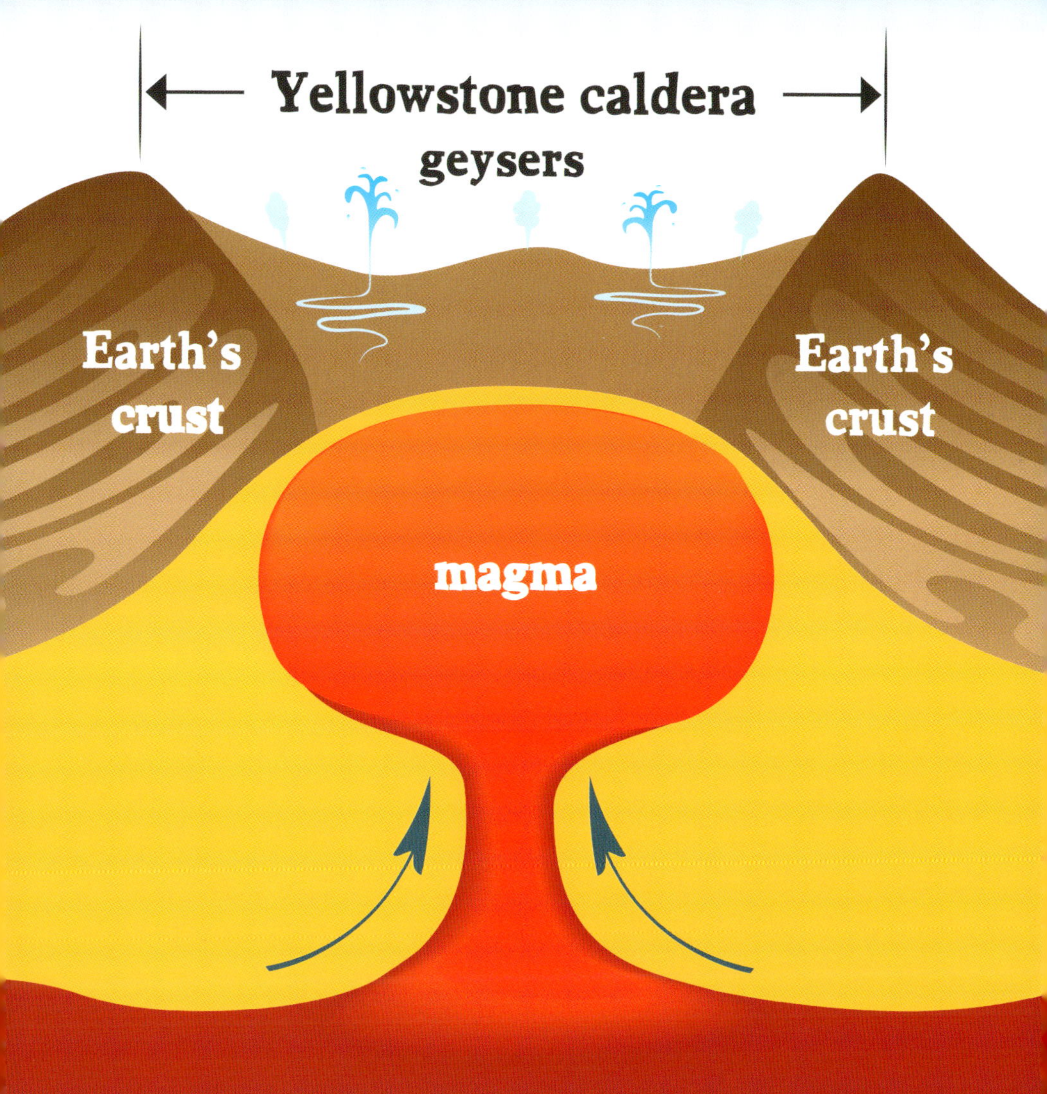

Earth has many tectonic plates that move. Sometimes they crash into each other and create a thin, narrow ditch.

For this to happen, scientists think a huge slab of an **oceanic plate** broke into pieces. A section of it sank deep into Earth. This section pushed hot materials upward into the **crust**. A supervolcano formed.

Earth has approximately 20 supervolcanoes. One created the massive caldera that is now Lake Toba in Indonesia.

The Yellowstone supervolcano hasn't erupted for 70,000 years. But there's still **magma** below the park. The magma

When magma goes onto Earth's surface, it is known as lava.

Grand Prismatic Spring is the biggest hot spring in Yellowstone.

heats layers of underground rock. Water from rain or melting snow sinks into the ground. When it comes into contact with these hot rocks, the water heats up. Then, the hot water rises to the surface. It can form a hot spring.

HEATING HOT SPRINGS

In some areas of Yellowstone, surface water sinks approximately 10,000 feet (3,048 m) into Earth. It's heated up by hot rocks before coming back to the surface and forming a hot spring. Geysers are a type of hot spring.

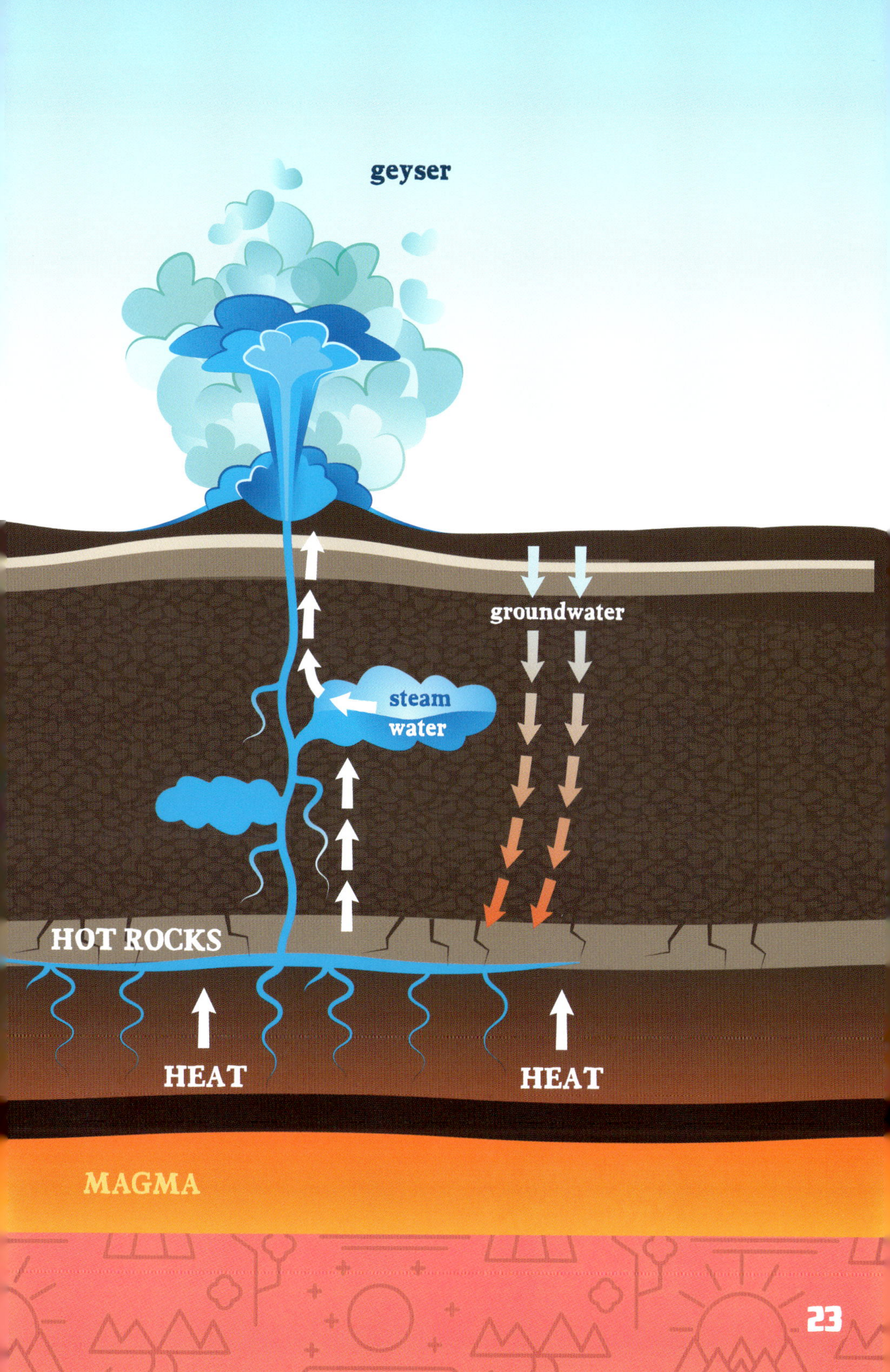
geyser
groundwater
steam
water
HOT ROCKS
HEAT
HEAT
MAGMA

CHAPTER 4

VISITORS AND THE CLIMATE

Researchers believe people have been soaking in Yellowstone's natural hot tub for many years. They found evidence that Native Americans lived near the river at one time. They also discovered that people had camped there in 1871.

COMPLETE AN ACTIVITY HERE!

Yellowstone became the country's first national park in 1872.

FUN FACT

Native Americans lived in the Yellowstone area long before it became a park. These groups included the Crow and the Umatilla. Today, the Crow Nation lives in Montana and the Umatilla live in Oregon.

However, the river's future is threatened. **Climate change** is causing

Warmer temperatures caused by climate change make it easier for wildfires to burn and spread in Yellowstone.

Climate change could make Yellowstone's geysers erupt less often.

less snowfall. Yellowstone's rivers and hot springs depend on slowly melting snow to refill the underground water systems. Without it, they may dry up.

Cars that run on gasoline release gases that pollute the air.

People can help slow down climate change. They can avoid littering and try to use other forms of transportation besides cars. These two things harm the environment. If people protect natural wonders such as Yellowstone's Boiling River, people will be able to enjoy the area for many more years.

People who visit the natural hot tub must be careful. There are no lifeguards to watch as they swim.

MAKING CONNECTIONS

TEXT-TO-SELF

If you visited Yellowstone National Park, would you want to soak in the natural hot tub? Why or why not?

TEXT-TO-TEXT

Have you read about other national parks? How were they similar to or different from Yellowstone National Park?

TEXT-TO-WORLD

Climate change could make some of the water features in Yellowstone dry up. How does climate change affect other parts of the world?

GLOSSARY

caldera — a large volcanic crater formed by a major eruption that caused the mouth of the volcano to collapse.

climate change — a crisis that is causing Earth's weather patterns to change, often including rising temperatures.

crust — the rocky, outer layer of Earth.

magma — rock beneath Earth's surface that has melted from intense heat.

oceanic plate — a huge piece of rock that sits at the bottom of the ocean and forms part of Earth's crust.

supervolcano — a large volcano that has had an explosive eruption that caused major effects on the climate and land.

INDEX

ONLINE RESOURCES

popbooksonline.com

Scan this code* and others like it while you read, or visit the website below to make this book pop!

popbooksonline.com/boiling-river

*Scanning QR codes requires a web-enabled smart device with a QR code reader app and a camera.